YOUR KNOWLEDGE HAS VALUE

- We will publish your bachelor's and
  master's thesis, essays and papers

- Your own eBook and book -
  sold worldwide in all relevant shops

- Earn money with each sale

Upload your text at www.GRIN.com
and publish for free

Air-Cooled Condenser Operation under Varying Ambient Conditions

Hans Georg Schrey

Bibliographic information published by the German National Library:

The German National Library lists this publication in the National Bibliography; detailed bibliographic data are available on the Internet at http://dnb.dnb.de.

ISBN: 9783963562136
This book is also available as an ebook.

© GRIN Publishing GmbH
Trappentreustraße 1
80339 München

Print and binding: Books on Demand GmbH, Norderstedt, Germany
Printed on acid-free paper from responsible sources.

The present work has been carefully prepared. Nevertheless, authors and publishers do not incur liability for the correctness of information, notes, links and advice as well as any printing errors.

GRIN web shop: https://www.grin.com/document/1450374

Title   **Operation of Air-Cooled Condensers under Varying Ambient Conditions**

Author   Hans Georg Schrey

Equations: Word equation editor

Figures: EXCEL/WORD graphics, created by author

On behalf of all authors, the corresponding author states that there is no conflict of interest. There are no other authors.

Keywords   Air-Cooled Condenser, ACC Design, Operation of ACC, Cross-Wind Effect, Forced Draft, Induced Draft

Abstract   Air-cooled condensers are widely used at the cold end of thermal power cycles. With this equipment cooling water required in wet cooled power plants may be saved on a large scale. To operate fans which convey dry cooling air to the condenser modules electric power is required which reduces overall power plant effectiveness. Consequently, pumping power consumption should be as small as possible. During operation it is therefore common to reduce fan power consumption by switching down or off fans of individual modules or, in some case complete ACC streets if ambient conditions and operation parameters are promising.

Performance guarantees are normally specified with all equipment in full action. Only primary variable variations such as ambient temperature, ambient air pressure or total steam flow rate for a reduced range at fan full speed are part of the performance guarantee. Therefore, effects of fan switch or high cross-wind levels remain open issues. The objective of the following report is to make up for these deficiencies at least on a general level – while still using first principles.

Contents

1 <u>Introduction</u>

As a consequence of increased environmental awareness dry air-cooled condensers (so-called ACC's) have gained wide acceptance in the realm of thermal power plants. Saving of cooling water is the main reason for this development.

In the past the investment for ACC's was nearly fourfold the cost of classical wet cooling systems. Lots of improvements over traditional technology have contributed to their success. Among other things, fin tube performance (multi row to single row) have been upgraded as well as steam side flow configuration (co-current, counter-current), steam side ducting and last not least fan design (high efficiency, low noise). Procedures for optimizing fin tube geometry as well as ACC design may be found as an example in quotation [1].

All these refinements have been made to reduce cost and to gain superior operation characteristics. To this end, it is indispensable to know the performance at varying operational conditions. ACC suppliers therefore have to provide performance diagrams at change of different primary variables - such as ambient air temperature, barometric pressure or steam side flow rate as a base framework.

General operational characteristics are important because design point conditions are never precisely met in practice. These characteristics are therefore paramount for optimizing the net energy generated by the power plant or to confirm contractual guarantees. The characteristics are commonly derived on the basis of first principles. This holds especially for corrections outlined in standard ACC acceptance test procedures [2], [3]. The purpose of these corrections is to estimate the difference between test and guarantee point performance.

Part load operation (fan or ACC street switch-off) is normally not part of the guarantee. Concerning cross-wind the allowable speed is limited for the guarantee [2], [3]. To assess the general effect of cross-wind CFD studies have been recommended. A detailed treatise of wind effects may be found in quotation [4], in particular §9.3 and §9.4. The recommendations there are primarily derived from on-site measurements of specific power stations. However, complex plant geometry with all of its buildings, landscape and environmental structures must be included into a comprehensive calculation model.

Commonly, power plant operators strive to put all uncertainties onto the ACC supplier by demanding guarantees for higher cross-wind velocities. This may result in considerable over-design as ACC manufacturers have the option of a "bold" or "conservative" cross-wind approach. However, neither a too costly investment nor an unfit cooling system is in the interest of the operator. This calls for the need of a general wind procedure.

Over the years, design cross-wind speed has been raised from 3 m/s to minimum 5 m/s at wind wall height as new standard. To make up for operational deficiencies caused by wind specific structures such as wind walls or over-design of circumferential fans have been proposed and successfully introduced in multiple installations. Three-dimensional air flow at cross-wind operation is meticulously covered in [4] and will not be repeated here. A detailed investigation on the effect of air side flow maldistribution in conventional two-pass ACC's may also be found in [5]. Unfortunately, the absolute maldistribution level remains an open issue.

To overcome all these theoretical shortcomings the topic of this report is to identify general rules - based on first principles - for the above-mentioned scenarios. The results will at least provide general trends for design considerations of ACC manufacturers or enable to assess secondary effects on the effective power plant performance.

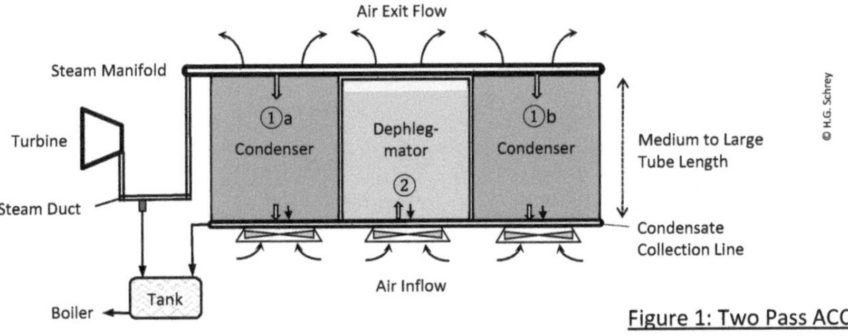

Figure 1: Two Pass ACC

2 ACC Layout

Traditionally, ACC's consist of two steam side flow passes: an A-frame arrangement of co-current condensation bundles, so-called *condensers*, and subsequently counter-current condensation bundles, so-called *dephlegmators*, equipped with forced draft fans (fig. 1). This technology has been in use for all plant sizes – from one-cell units to large cooling arrays consisting of multiple streets with multiple cells. In all types of air-cooled exchangers cross-wind flowing over and along the structure deteriorates cooling or condensation effectiveness and consequently, power plant output. ACC's are normally equipped with wind walls to protect the bundle exit area from direct wind impact. This allows a stable ACC operation - at least at low wind speed.

The idea of reducing cost as well as sensitivity to cross-wind brought about a new ACC variant comprising three steam side flow passes (dephlegmator – condenser – dephlegmator) equipped with induced fans, fig. 2. With three steam flow passes fin tubes must be shortened to avoid excessive steam side pressure drop and loss of performance. Typically, tube length is reduced to a level allowing cheap pre-manufacturing of complete modules in the workshop. Thus, on-site erection cost is reduced. With the induced fans mounted on top of the ACC the sensitivity to cross-wind is extremely reduced as well as hot air recirculation because of high exit air velocity in the fan bell. On the other hand, by enlarging the air exit speed by a factor of 3 or more and, at the same time increasing the volume flow rate caused by higher exit air temperature the air exit pressure loss is larger than in forced draft arrangements. As the cooling modules are arranged in an inverted A-frame layout the air inlet pressure drop however may be neglected.

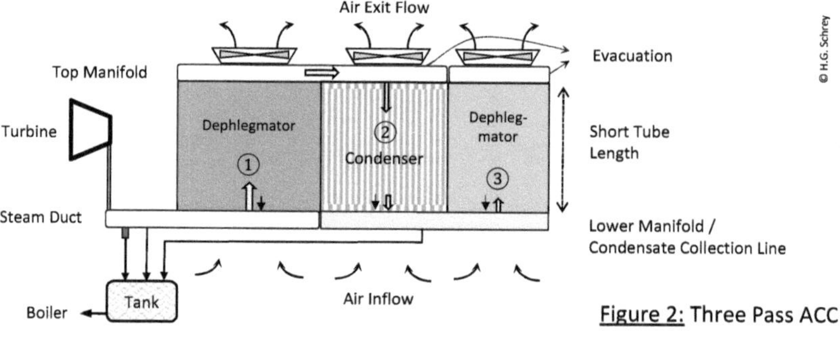

<u>Figure 2:</u> Three Pass ACC

In spite of the individual features of different types of ACC the general operational behavior may be summarized by first principle considerations. In this case there is no need for specific acceptance standards for different ACC arrangements.

The following procedure is based on calculation models outlined in contemporary acceptance test standards for air-cooled condensers. The general procedure is described in §5 of VGB-R 131 Me [2] and ASME PTC 30.1 [3]. However, extensions to the simple acceptance test models will be made to extend the validity range of the proposed equations to cover different fan settings. Two or three pass ACC's (fig. 1, 2) will be treated with the same procedures.

3   General ACC Model

In the following, for ease of overview lower case variables (temperature, pressure) refer to the steam side whereas upper case variables to the air side. A *module* is defined as the combination of heat transfer bundles (condenser or dephlegmator) cooled by one fan (fig. 1, 2). Moreover, tube bundle geometry is assumed to be the same in all modules, co-current or counter-current. From the thermal point of view the model will not distinguish condenser from dephlegmator modules which makes the procedure a lot easier.

Consequently, all ACC modules work on a similar level with same cooling characteristics. Thus, the general situation – irrespective of steam side arrangement – may be summarized as shown in figure 3 (see also [2]). The general ACC consists of a combination of air-cooled exchanger modules with connection lines and a turbine duct.

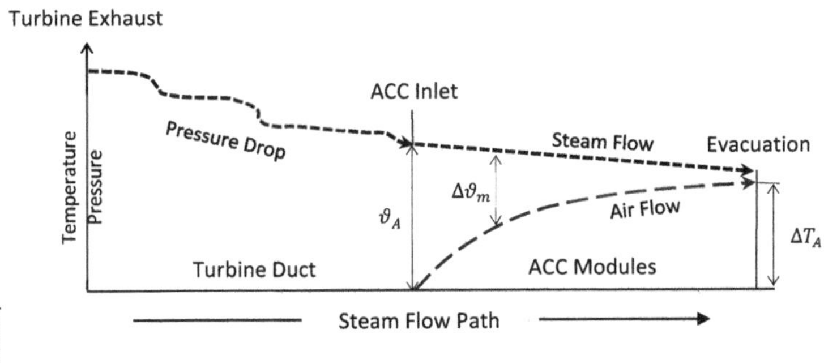

<u>Figure 3</u>: Temperature Profiles along the Steam Flow Path

Steam flowing through the duct is exposed to multiple local pressure losses along the flow path to the module section. The steam side pressure drop goes along with a reduction of steam temperature level (fig. 3) impairing module performance. Consequently, ACC suppliers strive to keep steam side pressure loss to a minimum. Within the modules steam flow encounters additional pressure loss until it vanishes at the terminal evacuation point. Condensate is extracted from the cycle via a collection line and drained to the condensate tank (fig. 1, 2). The terminal point of condensation is located at the end of the last dephlegmator bundle where the mixture of inert gas and remaining steam is extracted by the evacuation unit. At this point, the steam entering the turbine duct will be condensed completely. Steam quality decreases from x_0 at turbine exhaust to approximately $x_{ev} \cong 0$ at ACC evacuation point (fig. 3, 4). The small proportion of remaining steam contained in the terminal inert gas mixture can be disregarded from the perspective of total heat duty. A more detailed analysis of the physics involved may be found in quotations [6], [7] and [8] and will not be covered here.

Performance characteristics at variation of operational parameters are based on the design point. To use the simplified model all design point parameters must be defined based on this simplification. In the following all parameters at design point conditions will be denoted by index "0".

The core model is based on design steam side flow, quality, heat duty, turbine exhaust pressure, duct pressure loss as well as air side ambient temperature, barometric pressure and air temperature increase or, air volume flow rate respectively. Reference data are

Steam flow	$\dot{m}_0$
Turbine exhaust pressure	p_{EX0}
Condensation heat	Δh_{c0}
Steam quality	x_0
Air ambient pressure	B_0
Air ambient temperature	T_0
Air volume flow	$\dot{V}_{A0}$

With this simplification condensation heat r_0 and *steam quality* x_0 taken at *exhaust pressure* p_{EX0} suffice to find the design heat duty

$$\dot{Q}_0 = \dot{m}_0 \cdot \Delta h_{c0} \cdot x_0 \qquad (3\text{-}1)$$

This heat duty must be rejected by the ACC.

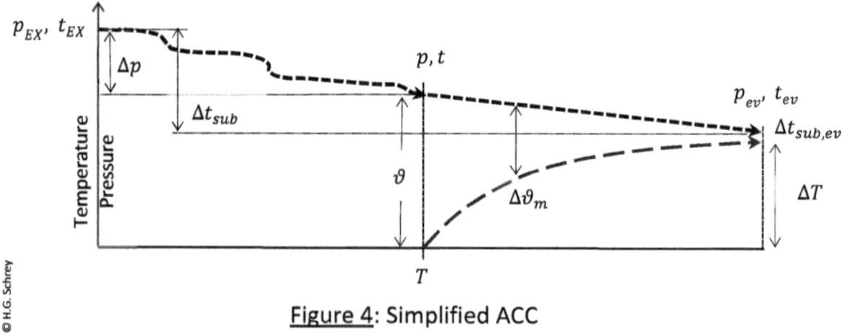

<u>Figure 4</u>: Simplified ACC

As for inter-dependency of steam pressure and steam temperature we shall use the *Antoine equation* for saturated steam. This is possible because the inert gas content plays only an extremely small role along the condensation path where the fraction of inert gas is negligible. Thus, for local steam temperature we only need local steam pressure (fig. 4).

The Antoine equation is
$$t_s = \frac{a}{b - \ln(p)} + c \qquad (3\text{-}2)$$

with constants $a = 3888.11$; $b = 23.308417$; $c = 43.2$ (all SI units)

The reverse function is
$$\ln(p) = b - \frac{a}{t_s - c} \qquad (3\text{-}3)$$

The *condensate subcooling level* Δt_{sub} (fig. 4) of contemporary ACC's is typically below 2 Kelvin ($\Delta t_{sub} \cong 1{,}5$ K). Therefore, the variation of steam temperature within the modules may be neglected and we shall restrict the operating conditions to the *module inlet state* (p, t) as dominant for the heat transfer process. Of course, base average heat transfer parameters must be defined accordingly.

Do not confuse Δt_{sub} with the *subcooling level of the gas mixture at dephlegmator outlet* $\Delta t_{sub,ev}$ which may be way higher (4K or more). This value is needed for the design of the evacuation unit. However, it plays no role in the thermal duty.

Steam side pressure drop depends on the geometry of steam ducting. Once the geometry is defined – we may easily find the module inlet state (p, t) by subtracting the *duct pressure loss* Δp from the turbine exit pressure p_{EX}. At design conditions we have

$$p_0 = p_{EX0} - \Delta p_0 \qquad (3\text{-}4)$$

All parameters are in SI units (K, Pa). This allows to define absolute temperature ratios

Steam side: $\qquad\qquad\qquad\qquad \tau \equiv t/t_0 \qquad\qquad\qquad\qquad (3\text{-}5)$

Air side: $\qquad\qquad\qquad\qquad \theta \equiv T_A/T_{A0} \qquad\qquad\qquad (3\text{-}6)$

At passing through the cooling bundles the steam side temperature encounters only a small decrease at nearly isothermal conditions. Therefore, we can use the isothermal exchanger effectiveness Φ which is solely dependent on the number of transfer units NTU:

$$\Phi = 1 - e^{-NTU} \qquad (3\text{-}7)$$

The *air temperature difference* is $\qquad\qquad \Delta T_A = \Phi\, \vartheta \qquad\qquad (3\text{-}8)$

and the *initial temperature difference* $\qquad \vartheta = (t - T_A) \qquad\qquad (3\text{-}9)$

For the design point we find $\qquad \Phi_0 = \dfrac{\Delta T_{A0}}{t_0 - T_{A0}} = \dfrac{\Delta T_{A0}}{\vartheta_0} \qquad\qquad (3\text{-}10)$

with $\qquad\qquad\qquad\qquad NTU_0 = -\ln(1 - \Phi_0) \qquad\qquad (3\text{-}11)$

Heat duty relates steam and air side parameters. The air side heat duty

$$\dot{Q} = c_A \cdot \rho_A \cdot \dot{V}_{A,tot} \cdot \Delta T_A \qquad (3\text{-}12)$$

is the same as steam side $\qquad \dot{Q} = (\dot{m} \cdot x) \cdot \Delta h_c \qquad (3\text{-}13)$

The total *air volume flow rate* $\dot{V}_{A,tot}$ depends on fan settings which – of course – are 100% speed at the design point. Fin tube geometry and fan type selection control the total level of air side flow rate.

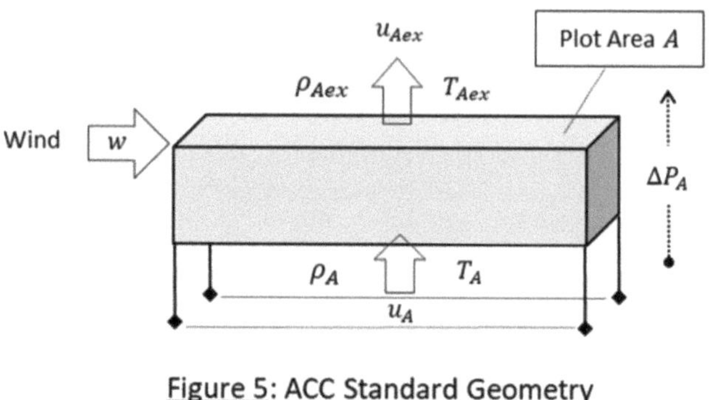

Figure 5: ACC Standard Geometry

Normally, heat exchanger geometry (and hydraulic length) varies from supplier to supplier. For the general calculation procedure a common parameter is needed. This suggests using *plot area A* which is defined in all types of ACC as the area through which the air flows. Thus, the heat exchange process can be expressed by a single average vertical air speed u_A (fig. 5) based on plot area A. This speed is also used for overall *air side pressure drop* Δp_A because the plot area does not change along the air side flow path. We can now replace

$$\dot{V}_{A,tot} = A_{tot} \cdot u_A \qquad (3\text{-}14)$$

The *total plot area* A_{tot} is a multiple of single *module plot areas* A_m. With the *total number of modules* n we find $\qquad A_{tot} = n \cdot A_m \qquad (3\text{-}15)$

and $\qquad \dot{V}_{A,tot} = n \cdot u_A \cdot A_m \qquad (3\text{-}16)$

if all fans are running at same speed n.

The *fan flow area* A_{fan} is different from *module plot area* A_m. This must be considered when evaluating the fan performance. Induced or forced draft arrangement are characterized by different fan volume flowrates. We define

$$\sigma_{mod} = A_m / A_{fan} \qquad (3\text{-}17)$$

This ratio is normally in the range $3 < \sigma_{mod} < 3{,}5$ for most ACC's. We therefore take the general value as $\sigma_{mod} \cong 3$.

Axial fans are air volume flow conveyors. Therefore, fan speed is directly proportional to air velocity. This assumption saves the complexity of incorporating fan performance curves which vary from type and supplier. With the *design average air speed* u_{A0} being known we find the actual air velocity based on *relative fan speed* s_i of each module "*i*"

$$u_{Ai} = s_i \cdot u_{A0} \tag{3-18}$$

as a first approximation

The *reference air speed* may easily be calculated via

$$u_{A0} = \dot{V}_{A,tot,0}/A_{tot} \tag{3-19}$$

If n_i fans are running at speed s_i the total volume flow is

$$\dot{V}_{A,tot} = A_m \cdot u_{A0} \cdot \Sigma_i(n_i\, s_i) \tag{3-20}$$

where

$$\Sigma_i(n_i) = n \tag{3-21}$$

The airside pumping power depends on air side pressure drop, fan type selection and number of modules. The *air side pressure drop* ΔP_A includes not only the fin tube loss but also all flow re-directions through the structure. In adverse circumstances also *wind effects* and hot *air recirculation* must be considered which we will come back later on.

With the air side pressure drop known the *total fan pumping power* for n fans is

$$\dot{N}_A = \frac{\dot{V}_{A,tot}\,\Delta P_A}{\eta_{tot}} \tag{3-22}$$

If fans are running at different speed the air side pressure head ΔP_{Ai} will change accordingly.

At different fan speed
$$\dot{N}_{A,s} = \frac{A_m\, u_{A0}}{\eta_{tot}} \cdot \Sigma_i(n_i\, s_i\, \Delta P_{Ai}) \tag{3-23}$$

where ΔP_{Ai} is a function of relative air velocity s_i.

The *overall fan efficiency* η_{tot} is composed of three contributions: *fan blade efficiency* η_F, *gearbox efficiency* η_{Gbx} and *motor efficiency* η_{Mot}

$$\eta_{tot} = \eta_F \cdot \eta_{Gbx} \cdot \eta_{Mot} \tag{3-24}$$

All efficiency factors are speed dependent. If we - simplifying - take the overall value η_{tot} constant we find no effect on the thermal process apart from the velocity variations.

The general heat balance is
$$\frac{\dot{Q}}{\dot{Q}_0} = \frac{\dot{m}\cdot x_{EX}}{\dot{m}_0\cdot x_{EX0}} \cdot \kappa_r = \frac{c_A\,\rho_A}{c_{A0}\,\rho_{A0}} \cdot \frac{A\cdot u_A}{A_{tot}\cdot u_{A0}} \cdot \frac{\Phi\,\vartheta}{\Phi_0\,\vartheta_0} \tag{3-25}$$

with *condensation heat ratio* $\kappa_r = \frac{\Delta h_c}{\Delta h_{c0}} \approx 1$ $\qquad\qquad$ (3-26)

In atmospheric cooling the variation of *air heat capacity* c_A is negligible. The *air density* follows the ideal gas law $\rho_A = B/(287{,}1 \cdot T)$ so that

$$\frac{\rho_A}{\rho_{A0}} = \frac{B\,T_{A0}}{B_0\,T_A} \qquad\qquad (3\text{-}27)$$

With $\qquad\qquad \beta \equiv B/B_0 \qquad\qquad (3\text{-}28)$

we find $\qquad\qquad \dfrac{c_A\,\rho_A}{c_{A0}\,\rho_{A0}} = \dfrac{\beta}{\theta} \qquad\qquad (3\text{-}29)$

Fans running at different speed go along with different *exchanger effectiveness* Φ_i.

Therefore, $\qquad \dfrac{\dot{V}_{A,tot}}{\dot{V}_{A,tot,0}} \cdot \dfrac{\Phi}{\Phi_0} = \dfrac{A_m \cdot u_{A0} \cdot \Sigma_i (n_i\,s_i) \cdot \Phi_i}{A_m \cdot u_{A0} \cdot n \cdot \Phi_0} = \cdot \Sigma_i \left(s_i\,\dfrac{n_i}{n} \cdot \dfrac{\Phi_i}{\Phi_0} \right)$

This turns (3-25) into $\qquad q = \dfrac{\dot{Q}}{\dot{Q}_0} = \dfrac{\beta}{\theta}\,\dfrac{\vartheta}{\vartheta_0} \cdot \Sigma_i \left(s_i\,\dfrac{n_i}{n} \cdot \dfrac{\Phi_i}{\Phi_0} \right) \qquad (3\text{-}30)$

Fan switch-off is included in (3-30). Even at switch-off a small natural draft contribution must be considered if passive modules remain open on the steam side. A good assumption for natural draft is $s_{ND} \approx 3\% - 7\%$ depending on wind wall height (the higher the larger).

Using the actual *initial temperature difference* ϑ we find the operating steam temperature

$$t = T + q\,\theta\,\frac{\vartheta_0}{\beta} \cdot \left[\Sigma_i \left(s_i \cdot \frac{n_i}{n} \cdot \frac{\Phi_i}{\Phi_0} \right) \right]^{-1} \qquad\qquad (3\text{-}31)$$

For the effectiveness Φ_i we use the approach similar to quotation [2]:

$$\left(\frac{\Phi_i}{\Phi_0} \right) = \frac{1 - e^{-NTU_i}}{1 - e^{-NTU_0}} \cong 1 - \Gamma + \Gamma \cdot \frac{NTU_i}{NTU_0} + \cdots \qquad (3\text{-}32)$$

with the auxiliary value $\qquad \Gamma \equiv \dfrac{NTU_0}{\exp(NTU_0) - 1} \qquad\qquad (3\text{-}33)$

This is good enough for relatively small changes of NTU. As practical design values are restricted to a narrow range of $1{,}1 < NTU < 1{,}6$ the linearization seems to be justified. Therefore, NTU can be used for basic design of air-cooled exchangers in general. A detailed analysis on the optimum design value may be found in quotation [9] for cross-counterflow arrangements which are standard in air coolers and condensers.

The variation of the effectiveness can now be reduced to changes of NTU being

$$\frac{NTU_i}{NTU_0} = \frac{\theta}{\beta\,s_i} \cdot \frac{U_i}{U_0} \qquad\qquad (3\text{-}34)$$

The *overall heat transfer coefficient* U_i is dominated by air side heat transfer. Tube wall and steam side contribute only a small proportion to total heat transfer resistance. *Fin side heat transfer* h_A follows the exponential of *air side Reynolds number Re* [2]

$$h_A \sim \lambda \cdot Re^{m_A} \sim \lambda \cdot \left(\frac{\rho}{\eta} u_A\right)^{m_A}$$

Consequently, we can assume a similar profile for overall U with a smaller exponent m. Based on the design point where *average air side heat transfer coefficient h_{A0}* and overall U_0 are known we can approximate the exponent by

$$m \cong m_A \cdot \left(\frac{U_0}{h_{A0}}\right) \tag{3-35}$$

For typical fin tube applications m_A is in the range of 0,3 to 0,45. The heat transfer coefficient ratio (U_0/h_{A0}) decreases with increasing exponent and vice versa. It is therefore justified to take an average value around $m \cong 0{,}3$.

For overall U we get

$$\frac{U_i}{U_0} \cong \kappa_{air} \cdot \left(\frac{\theta}{\beta s_i}\right)^{-m} \tag{3-36}$$

where

$$\kappa_{air} = \left(\frac{\lambda}{\lambda_0}\right) \cdot \left(\frac{\eta}{\eta_0}\right)^{-m} \cong 1 \tag{3-37}$$

This simplification holds for ACC's because the ambient air temperature is restricted by the maximum allowable steam temperature (imposed by turbine operation constraints).

The effectiveness is

$$\left(\frac{\Phi_i}{\Phi_0}\right) \cong 1 - \Gamma + \kappa_{air}\Gamma \cdot \left(\frac{\theta}{\beta s_i}\right)^{1-m}$$

The *steam temperature at module inlet* follows as

$$t \cong T_A + \frac{\theta\,\vartheta_0}{\beta} \cdot \frac{q}{S_\varphi} \tag{3-38}$$

with the definition

$$S_\varphi = \Sigma_i \left(s_i \frac{n_i}{n} \cdot \left[1 - \Gamma + \kappa_{air}\Gamma \left(\frac{\theta}{\beta s_i}\right)^{1-m}\right]\right) \tag{3-39}$$

Similarly, (3-30) turns into

$$q = \frac{\beta}{\theta}\left(\frac{\vartheta}{\vartheta_0}\right) S_\varphi \tag{3-40}$$

This equation combines all parameters relevant for heat duty.

The effective module pressure p can be calculated using temperature t (3-38) and the exponential of (3-3):

$$p = \exp\left(b - \frac{a}{t-c}\right) \tag{3-41}$$

4 ACC Steam Duct

We can now calculate from module inlet back to turbine exhaust by estimating the change of duct pressure drop Δp (fig. 4). Dry steam flow dominates the pressure drop. Keeping the duct resistance factor constant the *relative duct pressure loss* is

$$\frac{\Delta p}{\Delta p_0} = \frac{v_{st}}{v_{sto}} \frac{(\dot{m}\,x_{EX})^2}{(\dot{m}_0\,x_{EX0})^2} \tag{4-1}$$

with *mean specific steam volume v_{st}* between ACC inlet and turbine exhaust.

The equation of state in the steam duct combines specific volume, pressure and temperature. The duct is considered as isolated. Consequently, the steam temperature remains constant. Assuming that *relative duct mean pressure* is approximately the same as the *relative module pressure* we have

$$\frac{v_{st}}{v_{sto}} = \left(\frac{p}{p_0}\right)^{-1} \tag{4-2}$$

The *turbine exhaust pressure* is $\quad p_{EX} = p + \Delta p = p + \Delta p_0 \cdot \left(\frac{p}{p_0}\right)^{-1} \cdot \kappa_r^{-2} q^2 \tag{4-3}$

Generally, *steam duct design pressure drop* Δp_0 is restricted to a low percentage of the saturation pressure difference between turbine exhaust p_{EX0} and *saturation pressure at ambient air temperature* T_{A0} for which (3-41) yields

$$p_{A0} = \exp\left(b - \frac{a}{T_{A0} - c}\right) \tag{4-4}$$

Typically, the design pressure drop is selected as $\Delta p_0 \approx \kappa_P \cdot (p_{EX0} - p_{A0})$ with $\kappa_P = 5\%$.

Provided the design point data are known the calculation procedure is as follows:

a) Select ambient air temperature (T) and barometric pressure (B)
b) Calculate β (3-28) and θ (3-6)
c) Calculate ϑ_0 (3-9)
d) Select fan settings (n_i, s_i)
e) Select condensation heat ratio $\kappa_r \cong 1$ (3-26)
f) Select relative heat duty q (3-25)
g) Consider $\kappa_{air} = 1$ (3-37) and $m \cong 0,3$ (3-35)
h) Calculate Γ (3-33) using Φ_0 (3-10), NTU_0 (3-11)
i) Calculate S_φ (3-39)
j) Calculate module operating temperature t (3-38)
k) Calculate module operating pressure p (3-41)
l) Calculate turbine exhaust pressure p_{EX} (4-3)
m) Correct ratio κ_r (3-26) if necessary & repeat

Equations (3-38), (3-41) and (4-3) allow to set up performance charts for constant heat duty or constant air temperature but unfortunately, not for constant turbine exhaust pressure.

Examples of performance diagrams based on the above equations and procedure are shown in figures 6 and 7. Type 1 diagram (fig. 6) shows turbine exhaust pressure for constant ambient temperature as a function of relative heat duty.

Type 2 diagram (fig. 7) is an example for constant heat duty as a function of air temperature.

Both diagrams (as well as type 3 in chapter 5) are based on the design point:

$$p_{EX0} = 12{,}1 \text{ kPa}, \quad \kappa_r = 1, \quad x_{EX0} = 0{,}97 \text{ kg/kg}, \quad \Delta p_0 = 0{,}5 \text{ kPa}$$

$$B_0 = 720 \text{ mbar}, \quad T_0 = 20°C, \quad T_{ex0} = 39{,}9°C, \quad \text{Fans at Full Speed}$$

Figure 6: Performance Diagram Type 1

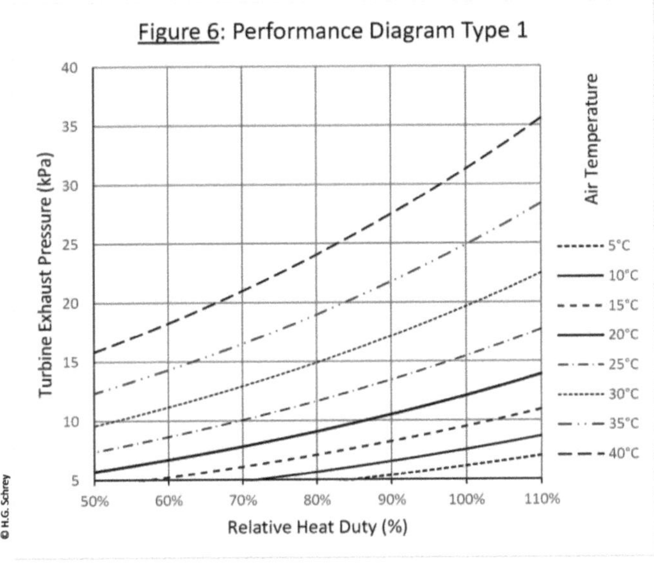

Figure 7: Performance Diagram Type 2

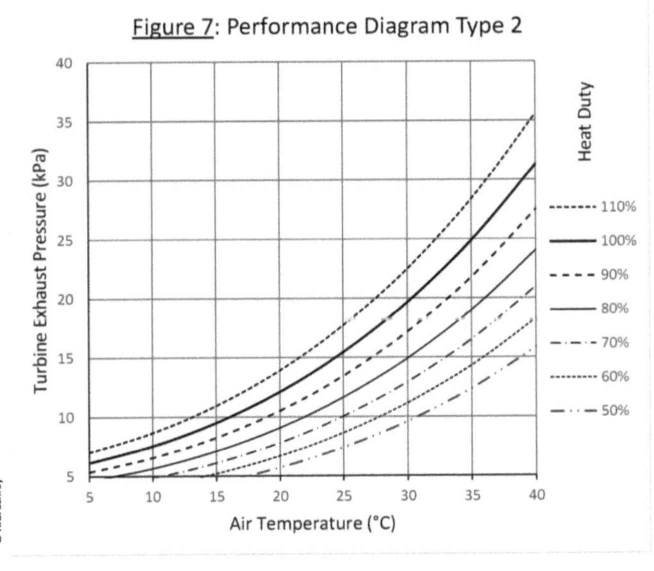

As we – so far – did not linearize the Antoine equation (3-41) the validity range of the proposed diagrams is extended as compared to [2] and [3]. Therefore, diagrams created by the above procedure may be used for general ACC performance.

5 Constant Exhaust Pressure

To arrive at a solution for constant turbine exhaust pressure the Antoine equation must be linearized. To this end, we define some parameter ratios.

- *Module pressure ratio* $\qquad \pi \equiv \left(\dfrac{p}{p_0}\right)$ $\hfill$ (5-1)

- *Exhaust pressure ratio* $\qquad \Pi \equiv \left(\dfrac{p_{EX}}{p_0}\right)$ $\hfill$ (5-2)

- *Duct pressure drop ratio* $\qquad \kappa_\Delta \equiv \left(\dfrac{\Delta p_0}{p_0}\right)$ $\hfill$ (5-3)

- *Antoine factor* $\qquad \kappa_0 \equiv \dfrac{a}{(t_0-c)^2}$ $\hfill$ (5-4)

- *Fan switch factor* $\qquad \kappa_q \equiv \dfrac{\beta S_\varphi}{\theta \vartheta_0}$ $\hfill$ (5-5)

At design point π and S_φ are equal to 1 while $\kappa_\Delta < 1$ and $\kappa_q = \vartheta_0^{-1}$. The solution comprises of three steps necessary to express exhaust pressure in function of heat duty. The 1st step relates module pressure to exhaust pressure according to (4-3).

Steam pressure drop (4-1) $\qquad \dfrac{\Delta p}{\Delta p_0} = \pi^{-1} \cdot \kappa_r^{-2} \, q^2$ $\hfill$ (5-6)

Exhaust pressure (4-3) $\qquad \Pi = \left(\dfrac{p_{EX}}{p_0}\right) = \pi + \pi^{-1} \cdot \kappa_\Delta \kappa_r^{-2} \, q^2$ $\hfill$ (5-7)

Note that module pressure $p\,[= \pi p_0]$ and temperature t are related via the saturation line. As p may deviate considerably from the design point (factor 2 or even more) a linearization of the Antoine equation must be done with caution. We therefore select steam temperature variation, typically between $10°C < t < 50°C$. This will be done in the 2nd step.

Consider, $\qquad \ln(p) = b - \dfrac{a}{\Delta+t_0-c} \cong b - \dfrac{a}{(t_0-c)} \cdot \left\{1 - \dfrac{\Delta}{(t_0-c)} + \cdots\right\}$

which is true as long as $\Delta = t - t_0 \ll (t_0 - c)$. The logarithm of p_0 is contained on the RHS, so that $\qquad \ln(\pi) \cong \dfrac{a\,t_0}{(t_0-c)^2}\,(\tau - 1) + \cdots$

Linearizing the exponent is possible because τ is close to 1. Therefore, using $\vartheta = (t - T_A)$ we find $\qquad \pi \cong \exp\big(\kappa_0 t_0 \cdot (\tau - 1)\big) \cong 1 + \kappa_0 t_0 \cdot (\tau - 1) + \cdots$

$$\pi \cong 1 - \kappa_0 t_0 + \kappa_0 t \qquad (5\text{-}8)$$

The steam temperature is $\qquad t \cong \dfrac{1}{\kappa_0} \cdot (\pi - 1 + \kappa_0 t_0)$ $\hfill$ (5-9)

In the 3rd step the heat duty function (3-40) comes into play. Inserting (5-9) into (3-40)

$$q \cong \frac{\beta S_\varphi}{\theta \vartheta_0}(t - T_A) = \frac{\kappa_q}{\kappa_0} \cdot (\pi - 1 + \kappa_0 t_0) - \kappa_q T_A$$

This may be re-arranged to
$$q + \frac{\kappa_q}{\kappa_0} - \kappa_q(t_0 - T_A) \cong \frac{\kappa_q}{\kappa_0} \cdot \pi \tag{5-10}$$

$$\pi \cong D + Cq \tag{5-11}$$

with
$$C \equiv \kappa_0/\kappa_q \tag{5-12}$$

$$D \equiv 1 - \kappa_0(t_0 - T_A) \tag{5-13}$$

As pressure ratio π is only an internal variable we re-arrange (5-10) for heat duty and insert into (5-7) to find the exhaust pressure as a function of heat duty q.

$$\Pi = (D + Cq) + \frac{\kappa_\Delta \kappa_r^{-2}}{(D+Cq)} \cdot q^2 \tag{5-14}$$

Note that equation (5-11) implies linearity between exhaust pressure and heat duty. This may be consistent only if q is close to 1. Re-arranging (5-14) to express heat duty as a function of exhaust pressure

$$\Pi \cdot (D + Cq) = (D + Cq)^2 + \kappa_\Delta \kappa_r^{-2} \cdot q^2$$

$$\Pi D + \Pi Cq = D^2 + 2CDq + C^2 q^2 + \kappa_\Delta \kappa_r^{-2} q^2$$

$$\Pi D - D^2 = 2q\, C(D - \Pi/2) + (C^2 + \kappa_\Delta \kappa_r^{-2})q^2$$

We find
$$q = \sqrt{E + F^2} - F \tag{5-15}$$

with
$$E \equiv \frac{\Pi D - D^2}{C^2 + \kappa_\Delta \kappa_r^{-2}} \tag{5-16}$$

$$F \equiv \frac{C(D - \Pi/2)}{C^2 + \kappa_\Delta \kappa_r^{-2}} \tag{5-17}$$

Provided the design point data are known the calculation procedure is as follows:

a) Select ambient air temperature (T) and barometric pressure (B)
b) Calculate β (3-28) and θ (3-6)
c) Calculate ϑ_0 (3-9)
d) Select fan settings (n_i, s_i)
e) Select condensation heat ratio $\kappa_r \approx 1$ (3-26)
f) Consider $\kappa_{air} = 1$ (3-37) and $m \approx 0{,}3$ (3-35)
g) Calculate Γ (3-33) using Φ_0 (3-10), NTU_0 (3-11)
h) Calculate S_φ (3-39) and κ_q (5-5)
i) Calculate Antoine factor κ_0 (5-4)
j) Calculate C (5-12) and D (5-13)
k) Select exhaust pressure p_{EX} and calculate Π (5-2)
l) Calculate E (5-16) and F (5-17)
m) Find relative heat duty q (5-15) and absolute value via (3-30)

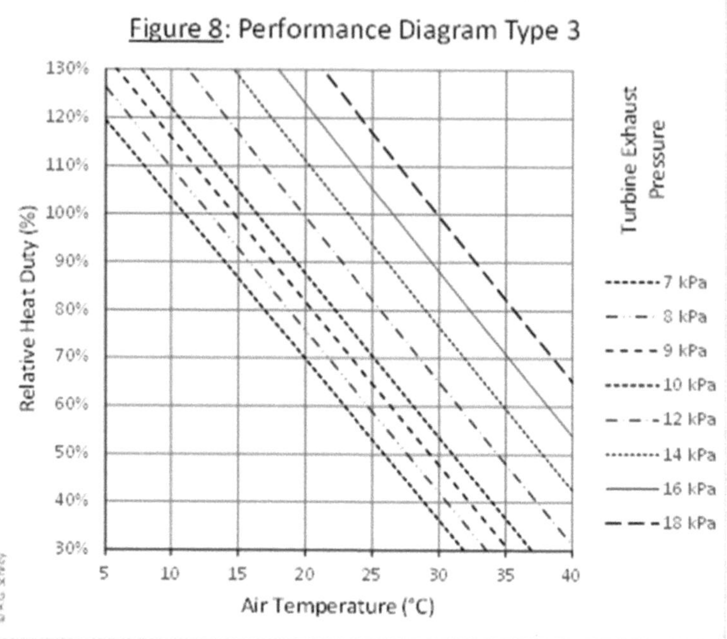

Figure 8 shows type 3 diagram for the same design data as in figures 6 and 7 (§4). As mentioned, because of the linearization the validity range of type 3 (fig. 8) is restricted. Nonetheless it may be used to assess the ACC performance within the acceptance window ($\pm$10% steam flow, $\pm$10K air temperature variation) as stipulated in codes [2] and [3].

6   Variation of Fan Speed

To estimate the effect of fan switches the dependency of air flow and fan speed is required. Basically, fan speed and air flow are assumed to be proportional – as long as the fans are the primary source of air flow. To avoid steam flow maldistribution fans are switched to the same level whenever possible. The fan speed in the example will be the same in all modules. The calculation procedure described in §3 and §4 will be used in the example. With the design data as stipulated in §4 (fig. 6 and 7) the turbine exhaust pressures at 80% and 60% relative fan speed are shown in figures 9 and 10.

As could be expected the turbine exhaust pressure p_{EX} rises with decreasing fan speed. As the Antoine equation has not been linearized the results may be used with confidence even at fan part load as long as speed variation is not excessive.

Note that in the case of full speed ($s = 1$) factor S_φ (3-39) reduces to

$$S_{\varphi,1} = 1 - \Gamma + \kappa_{air}\Gamma(\theta/\beta)^{1-m} \qquad (6\text{-}1)$$

Figure 9: Diagram Type 1 for 80% FanSpeed

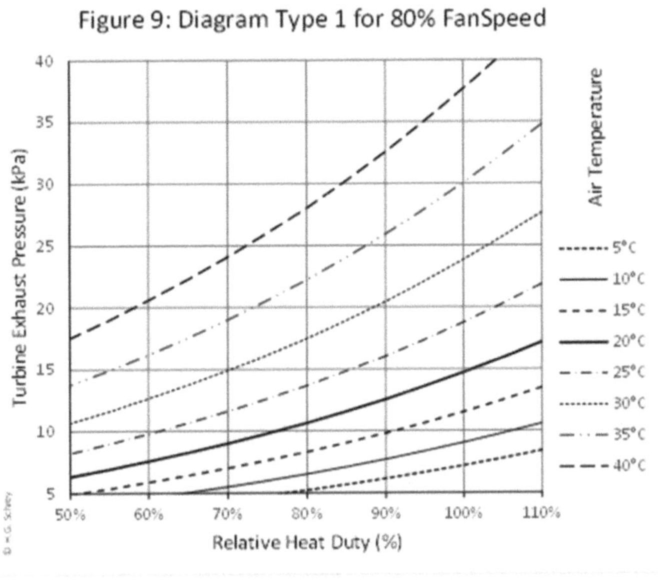

Figure 10: Diagram Type 1 for 60% FanSpeed

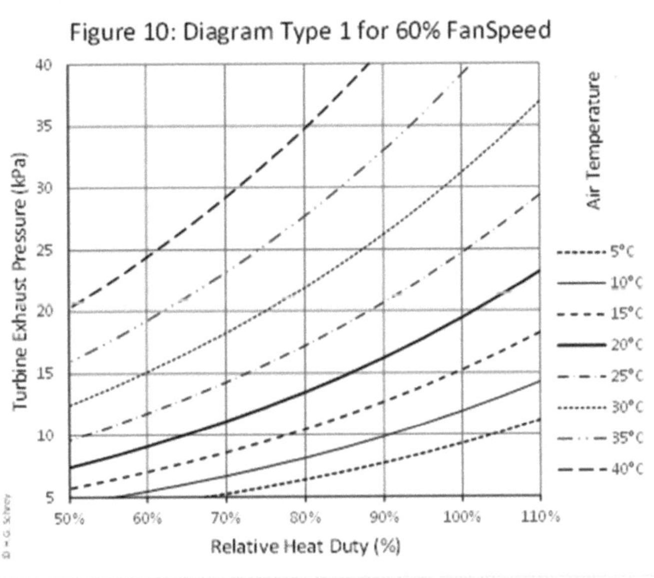

7   Cross Wind

As already mentioned cross-wind has a detrimental effect on the performance of air-cooled exchangers in general. Quotation [4] summarizes all effects involved and will not be repeated here. To reduce complexity the wind speed will be taken at top exit – valid for the whole ACC. Furthermore, the natural wind profile will be excluded from the considerations.

Mainly, there are three cross-wind effects (fig. 13):

  (1) Flow and temperature profile maldistribution

  (2) Additional air side pressure drop

  (3) Hot exit air recirculation

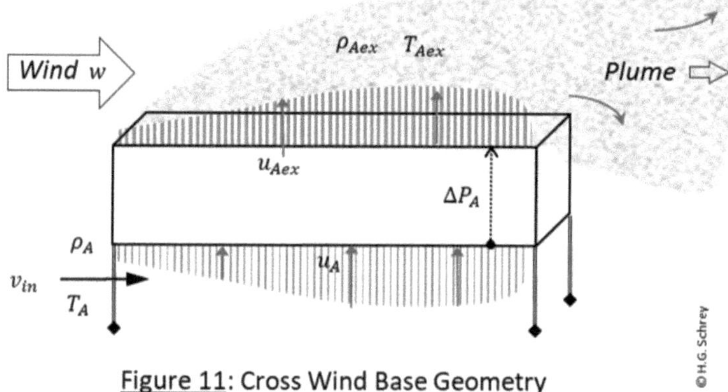

<u>Figure 11</u>: Cross Wind Base Geometry

(1) Flow and temperature profile maldistributions are local effects. Due to the three-dimensional air flow geometry it is impossible to assess the profile deflections on a general basis with undefined ACC geometry. For those who are interested please refer to §9 of quotation [4] where examples of the nature of air flow around the ACC at cross-wind conditions are meticulously described. In this situation we have to rely on procedures based on average values trusting that local effects are internally considered in the proposed solutions. Maldistribution effects cannot (and will not) be separated from either additional air side pressure loss or hot air recirculation. None-the-less, the procedures proposed will allow assessment of cross-wind effects - at least, on a basic level.

(2) To counteract the *additional air side pressure drop* Δp_W imposed by wind the fan design pressure drop Δp_A must be enlarged to prevent stall of fan air flow. Otherwise, the effective

air flow is reduced and heat duty goes down. With no detailed information available, we need a method to assess Δp_W.

(3) Increased cooling air temperature is the consequence of hot air recirculation. This reduces the effective initial temperature difference ϑ. Recirculation takes place not only downstream the wind flow direction but also along the sides of the ACC wind walls. This results in a lower initial temperature difference ϑ_0 which should be considered at the design stage. To this end, information on the level of recirculation is needed - preferably as a function of wind speed.

At planning stage, *wind walls* with appropriate *height* H_{ww} may be foreseen as well as external or internal wind shield devices or even overdesigned circumferential walkways to counteract negative effects. Even the support height of the ACC plays a role. It would be favorable to consider all the above effects in "as-built" condition – at least, to some extent.

In the following all cross-wind effects will be treated by using design point parameters. Results will be given in the form of relative variation of primary parameters. Because of the recirculation effect a small error in ambient temperature must be accepted. For nominal air inlet temperature equations (3-18), (3-39), (3-38), (5-8) and (4-3) reduce to

$$(3\text{-}18): \qquad s_W = u_A/u_{A0}$$

$$(3\text{-}39): \qquad S_\varphi \cong s_W \cdot \left[1 - \Gamma + \Gamma \left(\frac{\theta}{s_W}\right)^{1-m}\right]$$

$$(3\text{-}38): \qquad (t - t_0) \cong \vartheta_0 \cdot \left(\frac{\theta}{S_\varphi} - 1\right)$$

$$(5\text{-}8): \qquad \pi \cong 1 + \kappa_0(t - t_0)$$

$$(4\text{-}3): \qquad \Pi \cong \pi + \kappa_\Delta\, \pi^{-1}$$

where

$$\pi^{-1} \approx 1 - \kappa_0(t - t_0) + [\kappa_0(t - t_0)]^2 \pm \cdots$$

Note that the inverse of π has been linearized including the quadratic term of the temperature difference to enhance the precision of the results.

8 Cross-Wind – Pressure Drop

In the following we consider only average air velocities dominating the ACC flow field. The flow direction of hot exit air is bended by the impact of cross-wind (fig. 14) which redirects the exit flow momentum $\rho_{Aex} u_{Aex}$ by an *average deflection angle* α. Assuming that $u_W \approx u_{Aex}$ we find

$$\tan \alpha \approx \frac{\rho_A\, w}{\rho_{Aex} u_{Aex}} = \frac{w}{u_A} \qquad\qquad (8\text{-}1)$$

because of mass flow conservation, $\rho_{Aex} u_{Aex} = \rho_A u_A$. The flow momentum effect is mitigated to some extent by natural draft of the exit air flow. Furthermore, the average wind speed is presumably smaller than the impact speed w at ACC inlet. We therefore define the

Wind Number FD $$S_W \equiv \tan^{0,5}(\alpha) = \sqrt{w/u_A} \qquad (8\text{-}2)$$

where *FD* stands for "forced draft".

The *additional air side wind pressure drop* ΔP_W is dominated by wind number S_W. The higher the wind speed w the higher the additional pressure drop ΔP_W – and vice versa.

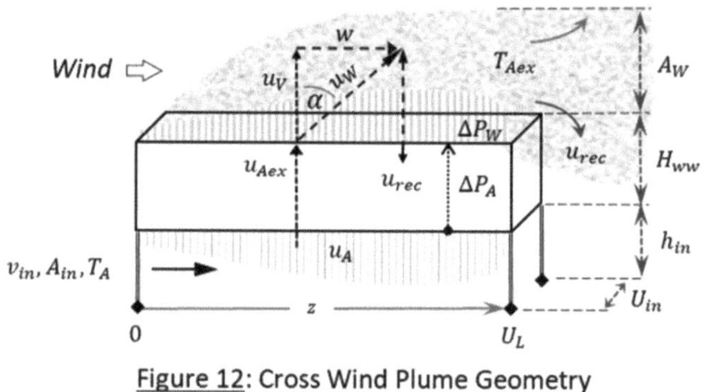

Figure 12: Cross Wind Plume Geometry

It is therefore logical to take the additional wind flow resistance as a function of wind speed:

$$\Delta P_W = C_W \frac{1}{2} \rho_A w^2 \qquad (8\text{-}3)$$

The *wind resistance factor* C_W is somewhere between 0 and 1. A good assumption for the general form of the *wind resistance factor* would be

$$C_W \cong \frac{S_W}{\xi + S_W} \qquad (8\text{-}4)$$

which starts at 0 and approaches 1 at extreme wind conditions. With no better information available we assume $\xi \cong 1$. Furthermore, we simplify by taking the average velocity at design point $u_A \rightarrow u_{A0}$ because variations of flow velocity u_A will be only on a small scale.

Thus, (8-2) takes the form $\qquad S_{W0} = \sqrt{w/u_{A0}} \qquad (8\text{-}5)$

and (8-4) $\qquad C_{W0} \cong \frac{S_{W0}}{1 + S_{W0}} \qquad (8\text{-}6)$

The wind resistance factor C_{W0} remains primarily a function of wind speed.

Additional air side pressure drop reduces the total air flow. Total fan power (3-22) is only marginally affected by pressure drop variations. This is due to the change of fan blade efficiency

(3-24). So, (3-22) cannot be used to estimate airflow variations. A general calculation model of fan airflow and static head is needed. Simplifying, fan performance at fixed blade angle may be approximated by an elliptical function. Figure 13 shows the general form.

We assume
$$\left(\frac{\dot{V}}{\dot{V}_{max}}\right)^2 + \left(\frac{\Delta P}{\Delta P_{max}}\right)^2 = 1$$

This curve must match the design point. Therefore,

$$\left(\frac{\dot{V}_0}{\dot{V}_{max}}\right)^2 + \left(\frac{\Delta P_{A0}}{\Delta P_{max}}\right)^2 = 1 \tag{8-7}$$

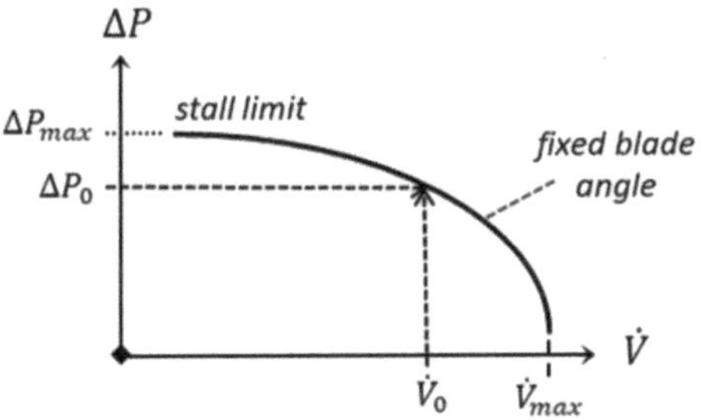

<u>Figure 13</u>: General Fan Performance Diagram

To avoid stall at high pressure head the fan design point should be at least 33% lower than theoretical maximum. This leads to the condition $\Delta P_{A0}/\Delta P_{max} \approx 67\%$ so that

$$\Delta P_{max} \cong 1{,}5\ \Delta P_{A0} \tag{8-8}$$

$$\dot{V}_{max} \cong 1{,}34\ \dot{V}_0 \tag{8-9}$$

and (8-7)
$$0{,}556 \cdot \left(\frac{\dot{V}}{\dot{V}_0}\right)^2 + 0{,}444 \cdot \left(\frac{\Delta P}{\Delta P_{A0}}\right)^2 = 1 \tag{8-10}$$

In this case $\dot{V}_0$ is the design fan volume flow and ΔP_{A0} the associated static head. This value corresponds to the airside pressure drop of the ACC. First of all, we define the ratios

Pressure drop
$$\varepsilon_P \equiv \frac{\Delta P}{\Delta P_{A0}} = \frac{\Delta P_{A0} + \Delta P_W}{\Delta P_{A0}} \tag{8-11}$$

Volume flow
$$\varepsilon_V \equiv \frac{\dot{V}}{\dot{V}_0} = \frac{u_A}{u_{A0}} \rightarrow s_W \tag{8-12}$$

The effective volume flow ratio ε_V may be interpreted as relative air speed s_W in equations (3-30) and (3-39).

If the enlarged pressure drop ΔP caused by further resistances is known we find from (8-10)

$$\varepsilon_P = 1{,}5 \cdot \sqrt{1 - 0{,}556 \cdot \varepsilon_V^2}$$

or, by linearizing
$$\varepsilon_P \cong 1{,}625 - 0{,}625\ \varepsilon_V^2 \tag{8-13}$$

As can be seen in (8-13) the relative operation point is independent from the fan arrangement as induced or forced draft fans share the same ε_P and ε_V. However, with induced fans the *initial exit flow velocity* is by factor σ_{mod} (3-17) larger than with forced fan arrangements. A modification of the air speed in (8-5) must be considered. Cross-wind meets areas of exiting airflow and spaces with no exit airflow which – in turn – means a reduction of average exit air speed. With no better information available we assume the *effective air speed* facing cross-wind at induced draft ID is

$$u_{A0,ind} \approx u_{A0}\sqrt{\sigma_{mod}} \qquad (8\text{-}14)$$

This leads to

$$S_{W0,ind} = \sqrt{w/u_{A0,ind}} \qquad (8\text{-}15)$$

and

$$C_{W0,ind} \cong \frac{S_{W0,ind}}{1+S_{W0,ind}} \qquad (8\text{-}16)$$

It is easy to see that (8-10) restricts the allowable additional pressure drop $(\Delta P_W/\Delta P_{A0})$ to 0,5. This is important because the linearized profile (8-13) may not be used above this limit. To evaluate (8-13) we need the un-affected pressure drop

$$\Delta P_A = \frac{1}{2}C_0\rho_A u_A^2 \qquad (8\text{-}17)$$

where

$$C_0 = \frac{2\,\Delta P_{A0}}{\rho_{A0}\,u_{A0}^2} \qquad (8\text{-}18)$$

Thus,

$$\frac{\Delta P_W}{\Delta P_A} = \frac{C_W\,w^2}{C_0\,u_A^2} = \frac{C_{W0}\,w^2}{C_0\,u_{A0}^2}\cdot \varepsilon_V^{-2}$$

and

$$\varepsilon_P - 1 = \frac{C_{W0}}{C_0}\cdot S_{W0}^4\cdot \varepsilon_V^{-2} \qquad (8\text{-}19)$$

With only small changes of air speed the inverse square expression in (8-19) can be linearized. We find

$$\varepsilon_V^{-2} \cong 1,5 + \varepsilon_V - 1,5\,\varepsilon_V^2$$

so that (8-19) reduces to

$$0,625 - 0,625\,\varepsilon_V^2 = \omega\,(1,5 + \varepsilon_V - 1,5\,\varepsilon_V^2)$$

We define

$$\omega \equiv \frac{C_{W0}}{C_0}\cdot S_{W0}^4 \qquad (8\text{-}20)$$

or, in case of induced fans

$$\omega_{ind} = \frac{C_{W0,ind}}{C_0}\cdot S_{W0,ind}^4 \qquad (8\text{-}21)$$

The relative air speed is

$$\varepsilon_V = \left[\frac{\omega}{3\omega-1,25}\right] + \sqrt{1 + \left[\frac{\omega}{3\omega-1,25}\right]^2} \qquad (8\text{-}22)$$

With ε_V known we get

$$S_\varphi \cong \varepsilon_V\cdot(1-\Gamma) + \Gamma\cdot\varepsilon_V{}^m$$

and using (3-38)

$$(t - t_0) \cong \frac{\vartheta_0}{S_\varphi} - \vartheta_0$$

we find

$$\left(\frac{p_{EX}}{p_0}\right) \cong (1+\kappa_\Delta) + (1-\kappa_\Delta)\cdot[\kappa_0(t-t_0)] + \kappa_\Delta[\kappa_0(t-t_0)]^2$$

where κ_Δ and κ_0 are taken from (5-3) and (5-4) respectively. The *relative change* of turbine exhaust pressure is

$$\left(\frac{p_{EX}}{p_{EX0}}\right) = \left(\frac{p_{EX}}{p_0}\right)\cdot\left(\frac{p_0}{p_{EX0}}\right) \qquad (8\text{-}23)$$

Figures 14 and 15 show results for the basic wind parameters at forced draft (8-5), (8-6) and induced draft (8-15) and (8-16). In case of induced fans the area ratio selected is $\sigma_{mod} = 3$, equation (3-17). As could be expected the wind parameters of induced draft are lower than those of forced draft arrangement. At 5 m/s (gentle breeze) the additional wind flow resistance factor C_{W0} is around 0,5 (forced) and 0,45 (induced).

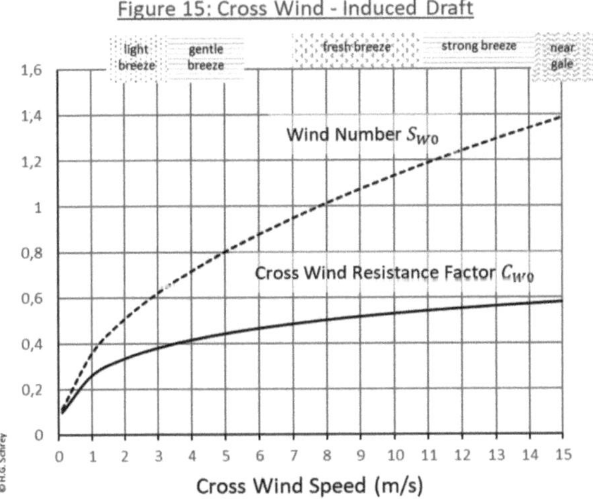

Figure 14: Cross Wind - Forced Draft

Figure 15: Cross Wind - Induced Draft

Gentle breeze is standard for contemporary design cases. This contribution should therefore be included in the fan design pressure drop. As can be seen resistance factors are in the range of $0,3 < C_{W0} < 0,6$ (low wind to strong breeze) for both induced and forced draft. This is confirmed by long-time experience.

Figures 16 and 17 show the exhaust pressure ratio (8-23) of both *FD* and <u>ID</u> configuration using the design data as listed in §4 for a typical range of airside design pressure drops.

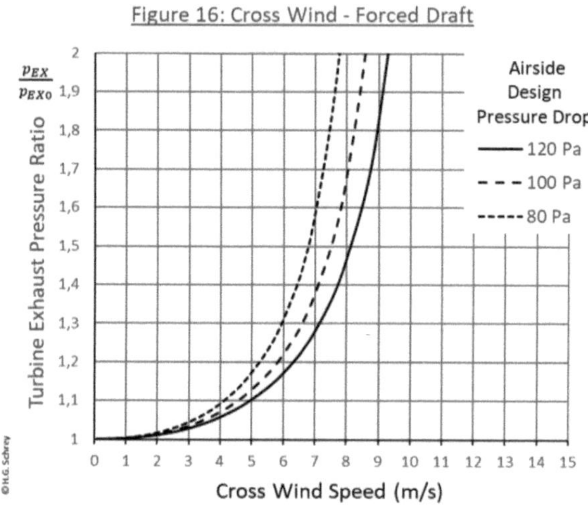

As expected, the design airside pressure drop influences the level of steam pressure variation. The higher the airside pressure drop the lower the effect on exhaust pressure and vice versa.

Also, the induced draft configuration (with area ratio $\sigma_{mod} = 3$) is much less susceptible to cross-wind than forced draft. In the range of low wind speed to gentle breeze the induced draft arrangement shows the same exhaust pressure effect only at 2 m/s higher cross-wind speed as compared to forced draft. However, this cross-wind advantage must be paid for by somewhat higher fan power as a result of enlarged air volume flow.

9 Cross-wind – Recirculation

Deflection angle α (fig. 12) may also be used to assess the air recirculation level. The vertical component of exit air speed u_V is approximately

$$[\rho_{Aex}]u_V = [\rho_{Aex}]u_W \cdot \cos \alpha \tag{9-1}$$

$$u_{rec} = f \cdot (u_{Aex} - u_V) \tag{9-2}$$

Only fraction f of $(u_{Aex} - u_V)$ will reach the air inlet h_{in} because the air is distributed into the ambient at downflow along the wind walls. From figure 18 we see that

$$\cos \alpha = u_V/u_W \cong u_V/u_{Aex}$$

so that
$$u_{rec} = u_{Aex} \cdot f\,(1 - \cos \alpha) \tag{9-3}$$

The fraction of exit airflow over the ACC wind wall top-down to the air inlet is the

recirculation rate
$$r \equiv u_{rec}/u_{Aex} = f\,(1 - \cos \alpha) \tag{9-4}$$

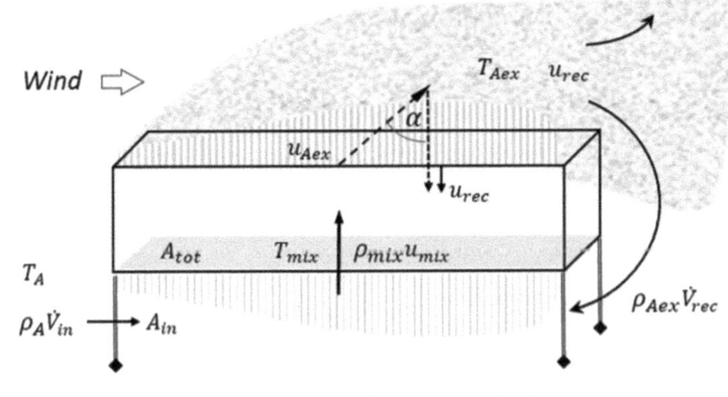

Figure 18: Cross Wind Recirculation

Note that in *ID* arrangement deflection angle (8-1) and wind number (8-2) must be calculated by using the average exit air speed $u_{A0,ind}$ as defined in equation (8-14).

A first assessment of the fraction finally reaching the air inlet is based on the length of the flow path top-down (fig. 19). As can be seen all air velocities are related to the plot area. The re-distribution of recircled exit air at downflow spans 3 dimensions. Therefore, we assume that only flow in one dimension (fig. 18) reaches the air inlet. This fraction is relevant for the *overflow length* H_{ww}^* of recirculation flow (fig. 19).

Thus, we define
$$f_r \approx \frac{1}{3}$$
(9-5)

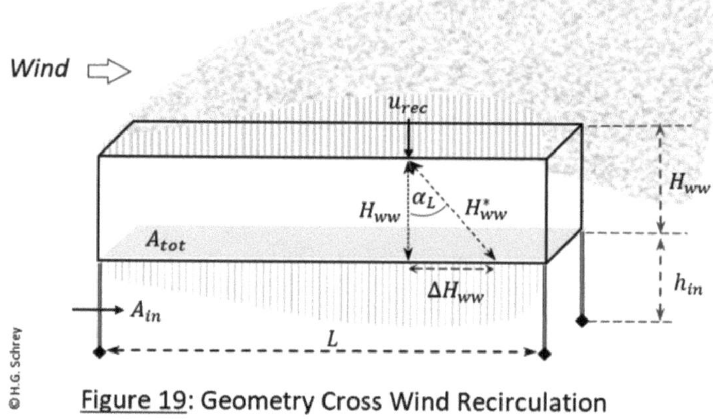

<u>Figure 19</u>: Geometry Cross Wind Recirculation

A broader insight into the characteristics of wind dominated air flow may be found in the measurement results quoted in [4]. In §9.4 (figures 9.4.4, 9.4.5, 9.4.6, 9.4.7 and 9.4.8) recirculation rates are generally measured in the range of $r < 10\%$. The proposed procedure must reflect these findings.

Figure 9.3.10 of [4] reveals that the ACC length in the direction of wind flow is dominant for recirculation. Considering figure 19 it is conclusive to assume a *longitudinal length ratio* as
$$\Lambda \approx \frac{L}{h_{in}+H_{ww}}$$
(9-6)

However, this holds only for unobstructed wind flow along the ACC wind wall. The longitudinal length L should be reduced otherwise.

At downflow of recirculated air the flow direction is deflected by the impact of wind speed in a similar way as at the ACC top (fig. 19). This *side deflection angle* α_L is similar to the overall deflection angle:
$$\alpha_L \approx \alpha$$
(9-7)

The *horizontal extension of wind wall height* can be seen on figure 19 as
$$\Delta H_{ww} \cong H_{ww} \tan \alpha$$
(9-8)

Thus, we find
$$H_{ww}^* \cong H_{ww}\,\sqrt{1+\tan^2\alpha} \qquad (9\text{-}9)$$

and *wind wall correction factor* $f_{ww} = \dfrac{h_{in}}{h_{in}+H_{ww}^*}$ (9-10)

The *total correction factor* is $\quad f = f_r\,\Lambda^{0,5}\,f_{ww}^{0,8}$ (9-11)

The exponents of the correction factors in (9-11) have been selected to avoid extreme figures and to match the measurement results of [4] as best as possible. Generally, the recirculation rate is calculated by equation (9-4).

At this point a distinction must be made between *FD* and *ID* geometry – see figure 20.

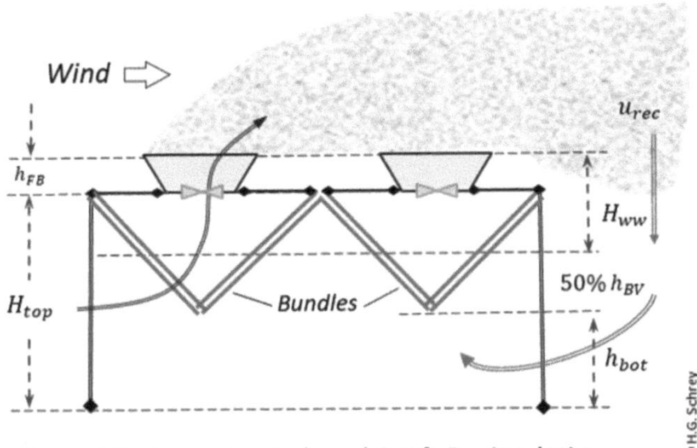

Figure 20: Geometry Induced Draft Recirculation

In *ID* there is no wind wall surrounding the bundles. Consequently, in *ID* some pseudo values must be introduced. The *narrowest gap for inlet flow* is the bottom distance between ground and bundle headers h_{bot}. This is interpreted as *induced inlet height*
$$h_{in,ind} = h_{bot} \qquad (9\text{-}12)$$
This value goes into (9-6) and (9-10) in case of *ID*. The *vertical bundle height* h_{BV} is connected to the *ACC top height* H_{top} as
$$h_{BV} = H_{top} - h_{bot} \qquad (9\text{-}13)$$
Typically, with reduced tube length we find $h_{BV} \le 4$ m where 50% of vertical bundle height is considered to be blocked for recirculation. Above the ACC top height there is the *fan bell and diffusor length* h_{FB} which is also considered blocked for recirculation.

Therefore, $\qquad H_{WW,ind} = \dfrac{1}{2}h_{BV} + h_{FB}$ (9-14)

In case of induced draft *ID* this value (9-14) goes into (9-6) and (9-10).

Figure 20: Recirculation Rate - Forced Draft
Longitudinal Length Effect

Figure 20 shows recirculation rate r (9-4) for various wind velocities and longitudinal length factors Λ in *FD* arrangement. Simplifying, the deflection angle α (8-1) was taken at 100% design speed u_{A0}. The profiles are similar to [4]. On figure 9.4.4 of quotation [4] recirculation profiles take a sharp increase with accelerating wind speed and go down after reaching a maximum somewhat above 10 m/s wind speed – depending on the wind approach angle. This is confirmed on figure 9.4.8 of [4]. The decrease of recirculation at extreme wind conditions is a result of wind pushing all air away from the ACC structure.

A similar situation arises in *ID* arrangements – see figure 21. The increase of recirculation rate with accelerated wind is smaller than in *FD* and reaches maximum values only above 15 m/s speed. As before, a *module plot to fan area ratio* (3-17) $\sigma_{mod} = 3$ was assumed.

Both recircled and fresh air are mixed at the exchanger inlet. Thus, the average temperature of the mixture follows from the heat balance

$$\rho_{mix}\dot{V}_{mix}\,c_P T_{mix} = \rho_A \dot{V}_{in}\,c_P T_A + \rho_{Aex}\dot{V}_{rec}\,c_P T_{Aex}$$

The continuity equation yields $\quad \rho_{mix}\dot{V}_{mix} = \rho_A \dot{V}_{in} + \rho_{Aex}\dot{V}_{rec}$

where

$$\rho_{mix}\dot{V}_{mix} = \rho_{mix}u_{mix}A_{tot} = \rho_{Aex}u_{Aex}A_{tot} \qquad (9\text{-}15)$$

$$\dot{V}_{in} = u_{in}A_{tot}$$

$$\dot{V}_{rec} = u_{rec}A_{tot}$$

This results in

$$\frac{\rho_A u_{in}}{\rho_{Aex}u_{Aex}} = 1 - r$$

Figure 21: Recirculation Rate - Induced Draft
Longitudinal Length Effect

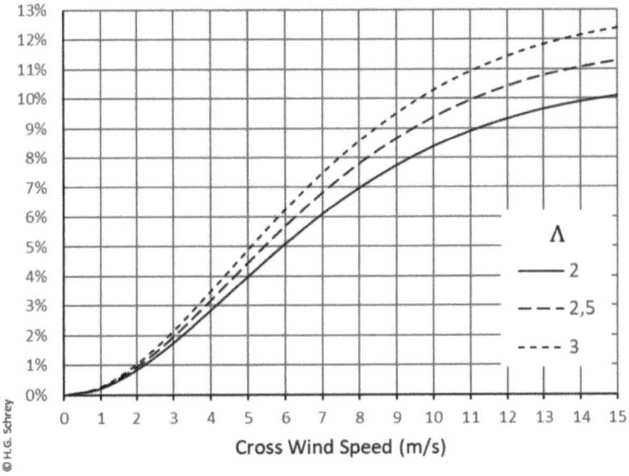

The heat balance may now be written as

$$\rho_{Aex} u_{Aex} A_{tot}\, c_P T_{mix} = \rho_A u_{in} A_{tot}\, c_P T_A + \rho_{Aex} u_{rec} A_{tot}\, c_P T_{Aex}$$

$$\rho_{Aex} u_{Aex} T_{mix} = \rho_A u_{in} T_A + \rho_{Aex} u_{rec} T_{Aex}$$

$$T_{mix} = (1 - r)\, T_A + r\, T_{Aex} \cong (1 - r)\, T_A + r\, (T_{mix} + \Delta T_A)$$

Simplifying, at $\Delta T_A \cong \Delta T_{A0}$ the *mixed air inlet temperature* will be

$$T_{mix} \cong T_A + \frac{r}{1-r}\, \Delta T_{A0} \qquad (9\text{-}16)$$

Note that in this scenario the *change of cooling temperature* $T_{mix} - T_A$ depends only on the design air temperature difference ΔT_{A0} (apart from the recirculation rate). Practically, the main influence on air temperature difference is caused by change of air volume flow. We conclude that the simplification made above match reality over the range of cooling air temperatures.

Examples of temperature change caused by mixing (9-16) are shown on figures 22 to 24 for forced draft and figures 25 to 27 for induced draft arrangements. Profiles are shown for different levels of design air temperature difference ΔT_{A0}. This agrees well with long-term experience of cooling tower performance under wind – compare figure 9.4.11 of quotation [4]. To ease comparison both *FD* and *ID* arrangements have been calculated for the same longitudinal length factors Λ. This means that the external ACC dimensions are not identical – which is no surprise.

Figure 22: Recirculation - Forced Draft

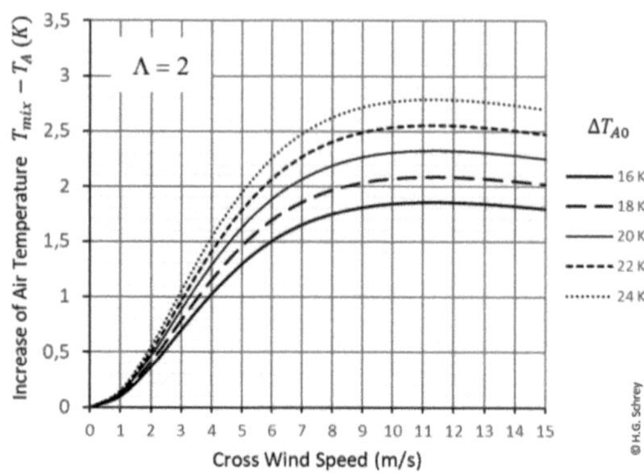

Figure 23: Recirculation - Forced Draft

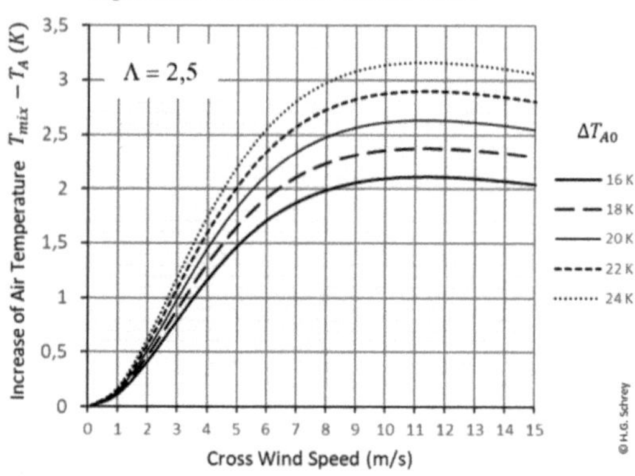

Figure 24: Recirculation - Forced Draft

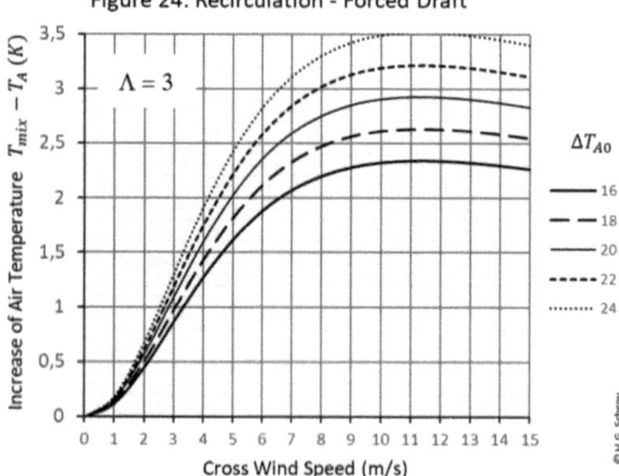

Figure 25: Recirculation - Induced Draft

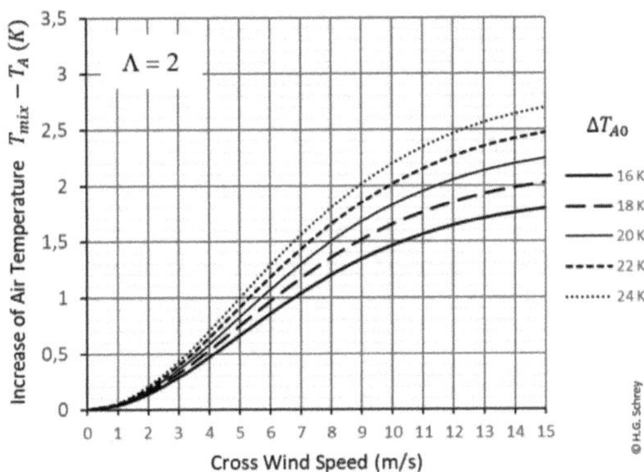

Figure 26: Recirculation - Induced Draft

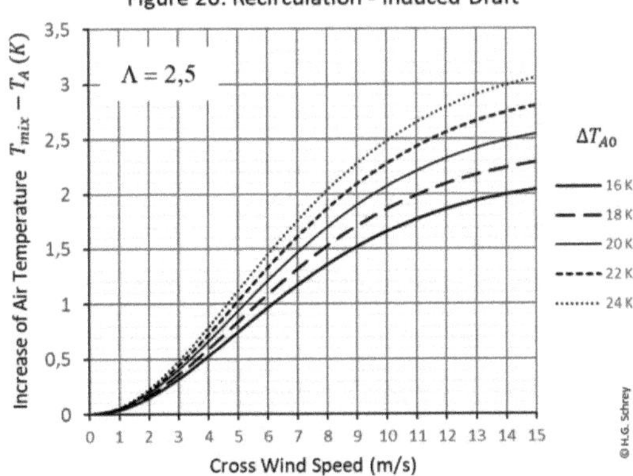

Figure 27: Recirculation - Induced Draft

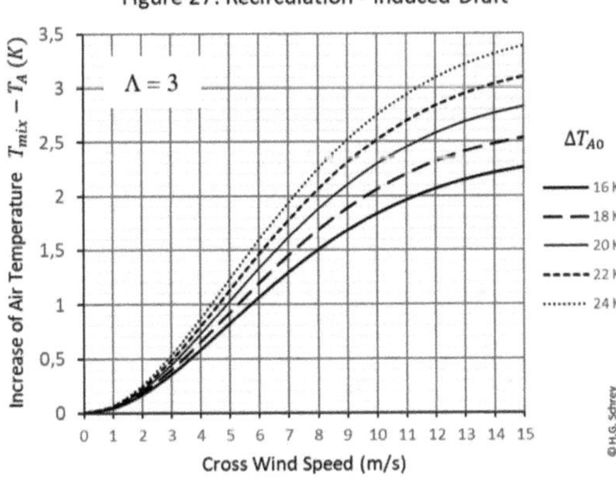

This information can be used in the standard performance diagrams (figures 6 to 8) by taking the results at mixed air inlet temperature. In forced draft arrangements the ambient temperature allowance is typically less than 2K at gentle breeze (5 m/s) while it is even less than 1K for induced draft as can be seen on figures 22 to 27.

10 Summary

The well-known mathematical model used in standard acceptance test procedures was upgraded to create diagrams incorporating general fan speed variations a well as fan switch-down or switch-off. Example results have been shown in different forms of diagrams.

In a further extension a simplified procedure has been derived to assess the effect of cross wind on the performance of the ACC within the range of the standard performance diagrams. With no "as built" data available it allows to approximate wind effects at design stage as well as at operation level.

The solution presented is based on first principles as well as long-term experience and measurements. For estimation of the effects only generally available data are required.

The procedure proposed in this report may be used at planning stage to counteract negative cross-wind effects and ensure that guarantees can be met even at higher wind speed. Based on the findings above the design ambient air temperature should be increased in the range of 1 to 1,5K. At the same time the estimated increase of air side pressure drop caused by wind should be considered as well because fan power consumption goes into the guarantee. Thus, a meaningful calculation can be made on the side of the ACC supplier. However, for the economic evaluation of the ACC design the percentage of expected annual working hours of high wind velocities must be considered.

Last not least, the wind effect equations may be used not only for ACC's but also for air cooler equipment in general provided the bundles are arranged in horizontal layout.

11 <u>Glossary</u>

All subscripts „0" refer to the design point

Index „i"   : number of module in ACC array (fig. 1, 2)

Index „m"   : single module

Index „ind"   : induced draft (fig. 2)

Index „rec"   : recirculated, recirculation

Superscript „*"   : changed value

Abbreviation FD : forced draft (fig. 1)

Abbreviation ID   : induced draft (fig. 2)

A_{tot} -   ACC plot area (m²) (3-15), (10-2)

A_m -   module plot area (m²) (3-15)

A_{fan} -   fan flow area (m²) (3-17)

B -   barometric pressure (Pa) (3-28)

C -   auxiliary factor (5-12)

D -   auxiliary factor (5-13)

E -   auxiliary factor (5-16)

F -   auxiliary factor (5-17)

c_A -   air specific heat capacity (J/kg K) (3-12)

λ -   air conductivity (W/m K) (3-37)

η -   air viscosity (m²/s) (3-37)

T_A -   ambient air temperature (K) (fig. 5, 11, 12)

ρ_A -   ambient air density (kg/m³) (3-27) (8-1)

T_{Aex} -   exit air temperature (K) (fig. 5, 11, 12)

ρ_{Aex} -   exit air density (kg/m³) (8-1)

ΔT_A -   air temperature difference (K) (3-8)

$\dot{V}_{A,tot}$ -   air volume flowrate (m³/s) (3-14), (3-16), (3-20)

ΔP_A -   airside pressure drop (Pa) (8-17) (fig. 12)

s_i -   relative air speed, speed of fan no. "i" (3-18)

n_i -   number of fans "i" running at speed s_i (3-21)

n -   total number of modules (3-21)

Φ -   exchanger effectiveness (3-10)

NTU -   number of transfer units (3-11)

Γ - auxiliary value exchanger effectiveness (3-33)

h_A - fin side heat transfer coefficient (W/m²K) (3-35)

m_A - fin side heat transfer Reynolds exponent (3-35)

U - overall heat transfer coefficient (W/m²K) (3-36)

u_A - vertical air speed (m/s) (3-19) (fig. 5, 11, 12)

$u_{A0,\text{ind}}$ - design vertical air speed (m/s), induced draft (8-14)

m - average overall U Reynolds exponent (3-35)

S_φ - effectiveness auxiliary value (3-39)

$S_{\varphi,1}$ - effectiveness auxiliary value, full speed case (6-1)

κ_q - fan switch factor (5-5)

t - steam temperature (K) at module inlet (3-5), (3-2), (3-38)

t_s - saturated steam temperature (K) (3-2)

Δ - temperature difference (K) at linearization (§5)

κ_0 - Antoine factor (5-4)

ϑ - initial temperature difference steam - air (K) (3-9)

$\dot{m}$ - turbine exhaust steam flow (kg/s) (3-13)

p - module inlet steam pressure (Pa) (3-4), (3-41)

p_{EX} - turbine exhaust steam pressure (Pa) (3-4), (4-3)

Δp - turbine duct pressure drop (Pa) (3-4), (4-1)

$\dot{Q}$ - heat duty (W) (3-12), (3-13)

q - heat duty ratio (3-30), (3-40)

Re - airside Reynolds number

Δh_c - condensation heat (J/kg) (3-26)

v_{st} - mean steam duct specific volume (m³/kg) (4-2)

x - steam quality (kg/kg) (3-13)

β - air barometric pressure ratio (3-28)

κ_{air} - airside property ratio (3-37)

κ_r - condensation heat ratio (3-26)

κ_Δ - duct pressure drop ratio (5-3)

σ_{mod} - module plot to fan area ratio (3-17)

θ - ambient air temperature ratio (3-6)

τ	-	module steam temperature ratio (3-5)
π	-	module pressure ratio (5-1)
Π	-	exhaust pressure ratio (5-2)

Wind procedure parameters:

u_{rec}	-	vertical recirculation air speed (m/s) (9-3) (fig. 18)
α, α_L	-	exit air deflection angle (8-1), side deflection angle (9-7) (fig. 12, 18, 19)
ε_P	-	air pressure drop ratio (8-11)
ε_V	-	air volume flow ratio (8-12)
ω	-	auxiliary wind resistance factor (8-20), (8-21)
S_W	-	wind number (8-2), (8-5)
C_W	-	wind pressure drop resistance factor (8-4)
ΔP_W	-	additional pressure drop caused by cross-wind (Pa) (8-3)
ΔP_{max}	-	maximum fan pressure head (Pa) (8-7)
$\dot{V}_{max}$	-	fan air volume flow limit (m³/s) (8-7)
C_0	-	average air pressure drop coefficient (8-18)
h_{in}, h_{bot}	-	inlet (support or bottom) height (m) (fig. 12, 19, 20)
h_{BV}	-	vertical bundle height (m) (fig. 20)
h_{FB}	-	fan bell and diffusor height (m) (fig. 20)
H_{WW}	-	wind wall height (m) (fig. 12, 19)
H_{top}	-	vertical height of ACC (m) (fig. 20)
ΔH_{WW}	-	wind wall height (m) (9-8) (fig. 19)
H_{WW}^*	-	recirculation overflow length (m) (9-9) (fig. 19)
f	-	recirculation rate factor (9-11)
f_r	-	recirculation factor (9-5)
f_{ww}	-	wind wall correction factor (9-10)
L	-	length of wind direction (m) (9-6) (fig. 19)
Λ	-	longitudinal length correction factor (9-6)
r	-	air recirculation rate (9-4)
T_{mix}	-	average mixed air temperature at inlet (K) (9-16)
u_{mix}	-	mixed air velocity at inlet (m/s) (9-15) (fig. 19)
ρ_{mix}	-	mixed air density at inlet (kg/m³) (9-15)

12 <u>Bibliography</u>

[1]   Schrey, H.G.: „Thermohydraulic Comparison of Fin Tubes", München, GRIN Verlag
(2017), https://www.grin.com/document/414394

[2]   "VGB Guideline Acceptance Test Measurements and Operation Monitoring of Air-
Cooled Condensers under Vacuum", VGB R-131 M e, published by "VGB Technische
Vereinigung der Grosskraftwerksbetreiber", 1997

[3]   "Air-Cooled Steam Condensers, Performance Test Codes", ASME PTC 30.1 - 2007, The
American Society of Mechanical Engineers, 2008

[4]   Detlev G. Kröger (Department of Mechanical Engineering, University of Stellenbosch)
"Air-Cooled Heat Exchangers and Cooling Towers", Begell House (1998)

[5]   Schrey, H.G.: „Effect of Uneven Cooling on Performance of Air-cooled Condenser",
München, GRIN Verlag (2017), https://www.grin.com/document/418540

[6]   Schrey, H.G.: „Review of the Vacuum Decay Test in Air-cooled Steam Condensers",
München, GRIN Verlag (2018), https://www.grin.com/document/455056

[7]   Schrey, H.G.: „Flow Maldistribution in Tube Bundles. Application on Air-cooled Steam
Condensers", München, GRIN Verlag (2018), https://www.grin.com/document/432886

[8]   Schrey, H.G.: „Vacuum Drop Test of Air-cooled Condensers in Operation", München,
GRIN Verlag (2018), https://www.grin.com/document/421324

[9]   Schrey, H.G.: "Practical Aspects of Air-Cooled Exchanger Design", 12th IAHR Cooling
Tower & Spraying Pond Symposium 2001, UTS Sydney, Australia